Global Issues
Toxic Waste

Cheryl Jakab

Smart Apple Media

This edition first published in 2010 in the United States of America by Smart Apple Media.
All rights reserved. No part of this book may be reproduced in any form or by any means without written permission from the publisher.

Smart Apple Media
P.O. Box 3263
Mankato, MN 56002

First published in 2009 by
MACMILLAN EDUCATION AUSTRALIA PTY LTD
15–19 Claremont Street, South Yarra, Australia 3141

Visit our Web site at www.macmillan.com.au or go directly to www.macmillanlibrary.com.au

Associated companies and representatives throughout the world.

Copyright © Cheryl Jakab 2009

Library of Congress Cataloging-in-Publication Data

Jakab, Cheryl.
    Toxic waste / Cheryl Jakab.
        p. cm. – (Global issues)
    Includes index.
    ISBN 978-1-59920-455-0 (hardcover)
    1. Hazardous wastes–Juvenile literature. 2. Hazardous substances–Juvenile literature. I. Title.
    TD1030.5.J34 2010
    363.72'87–dc22
                                        2009002024

Edited by Julia Carlomagno
Text and cover design by Cristina Neri, Canary Graphic Design
Page layout by Christine Deering and Domenic Lauricella
Photo research by Jes Senbergs

Printed in the United States

Acknowledgments
The author and the publisher are grateful to the following for permission to reproduce copyright material:

Front cover photograph: Chemical waste running into river, photo by David Woodfall/Getty Images

Photos courtesy of: AP Photo/AAP, 7 (top left), 7 (top right), 13, 17; Rob Cruse, 25, 27, 29; AFP/Getty Images, 15, 23, 12; Getty Images, 6 (right), 21, 18, 19; © Floortje/Istockphoto, 7 (bottom), 24; © Wolfgang Wiedemann/istockphoto, 5; Photolibrary, 16; Photolibrary/ © Barbimages/Alamy, 22; Photolibrary/Dr Jeremy Burgess/Science Photo Library, 14; Photolibrary/ © Directphoto.org/Alamy, 26; Photolibrary/ © D. Hurst/Alamy, 11; Photolibrary/ © Photofusion Photo Library/Alamy, 6 (left), 9; Photolibrary/ © Stock Connection Blue/Alamy, 10; Photolibrary/ © Visions of America, LLC/Alamy, 20.

While every care has been taken to trace and acknowledge copyright, the publisher tenders their apologies for any accidental infringement where copyright has proved untraceable. Where the attempt has been unsuccessful, the publisher welcomes information that would redress the situation.

Please note
At the time of printing, the Internet addresses appearing in this book were correct. Owing to the dynamic nature of the Internet, however, we cannot guarantee that all these addresses will remain correct.

Contents

Facing Global Issues   4

What's the Issue? Toxic Waste   5

Toxic Waste Issues Around the Globe   6

ISSUE 1
Exposure to Toxic Organic Chemicals   8

ISSUE 2
Lead Poisoning   12

ISSUE 3
Mercury Poisoning   16

ISSUE 4
Stockpiles of Toxic Chemicals   20

ISSUE 5
Toxic Chemicals at Home   24

What Can You Do? Choose Non-toxic Products   28

Toward a Sustainable Future   30

Web Sites   30

Glossary   31

Index   32

Glossary Words
When a word is printed in **bold**, you can look up its meaning in the Glossary on page 31.

Facing Global Issues

Hi there! This is Earth speaking. Will you spare a moment to listen to me? I have some very important things to discuss.

We must face up to some urgent environmental problems! All living things depend on my environment, but the way you humans are living at the moment, I will not be able to keep looking after you.

The issues I am worried about are:
- the effects of **global warming**
- the health of natural environments
- the use of **nonrenewable** energy supplies
- the environmental impact of unsustainable cities
- the build-up of toxic waste in the environment
- a reliable water supply for all

My global challenge to you is to find a sustainable way of living. Read on to find out what people around the world are doing to try to help.

Fast Fact
Sustainable development is a form of growth that lets us meet our present needs while leaving resources for future generations to meet their needs too.

What's the Issue?
Toxic Waste

Today, toxic waste is building up in the environment. It is poisoning and even killing living things. Toxic waste is the cause of serious health problems for many people.

What is Toxic Waste?

Toxic waste is dangerous chemicals that have been set aside as waste. Toxic waste includes:
- **organic chemicals**, such as many **pesticides**
- **inorganic chemicals**, such as lead and mercury
- **radioactive materials**, such as those used in nuclear power plants, nuclear weapons, and medicine

Fast Fact
Toxic chemicals can enter the body through the nose, mouth, or skin.

Sources of Toxic Waste

Sources of toxic waste include industry, agriculture, and households. Many household substances, such as cleaning products, contain toxic chemicals. The chemicals may be **poisons** or they may become toxic after being **exposed** for a long time. These chemicals can become toxic waste if they are not disposed of properly.

Many people are exposed to toxic waste when they handle dangerous chemicals. People are also suffering from the effects of exposure to **chemical residues** all over the globe. Plants and animals are exposed to toxic waste when toxic chemicals are released into the environment.

Some factories produce toxic waste by burning chemicals and releasing poisons into the air.

Toxic Waste Issues

Around the globe, issues associated with toxic waste include:
- exposure to toxic organic chemicals (see issue 1)
- people being poisoned by lead (see issue 2)
- people being poisoned by mercury (see issue 3)
- stockpiles of toxic chemicals causing chemical spills (see issue 4)
- children being poisoned at home (see issue 5)

ISSUE 1

United States
Local birds died after being exposed to the pesticide DDT. See pages 8–11.

ISSUE 4

Bhopal, India
Thousands died and many were injured following a chemical spill in Bhopal. See pages 20–23.

Around the Globe

ISSUE 2

China
People are poisoned by lead from electronic waste.
See pages 12–15.

ISSUE 3

Minamata, Japan
People have been poisoned by mercury pollution in Minamata Bay.
See pages 16–19.

ISSUE 5

Australia
Children are poisoned by toxic chemicals at home.
See pages 24–27.

ISSUE 1

Exposure to Toxic Organic Chemicals

Many people and animals are exposed to toxic organic chemicals because the chemicals are not known to be dangerous. Exposure to toxic organic chemicals can cause environmental diseases.

What are Toxic Organic Chemicals?

Organic chemicals are complex chemicals formed by living things and containing carbon. Some common organic chemicals are listed in the table below.

Toxic Organic Chemicals
benzene
chloroform
dichlorodiphenyltrichloroethane (DDT)
polychlorinated biphenyls (PCBs)
vinyl chloride

Use of Toxic Organic Chemicals

Many toxic organic chemicals were widely used before they were known to be dangerous. Manufactured organic compounds, such as DDT and polyvinyl chloride (PVC), have been used in industry and agriculture since the early 1900s. They were used as pesticides, in plastics and in many other products. Today, the chemical residues of these compounds **contaminate** areas all over the globe.

Environmental Diseases

Environmental diseases are illnesses caused by exposure to toxic chemicals in the environment. Every year, environmental diseases are recorded in each country. One important group of environmental diseases are occupational illnesses. These are illnesses caused by exposure to toxic substances in the workplace.

Fast Fact
Toxic organic chemicals can be dangerous to humans. Exposure to vinyl chloride and benzene can cause cancers. PCBs can affect brain development and mental ability in children.

CASE STUDY
Birds Poisoned by DDT in the United States

DDT was an organic pesticide first used in the United States in the 1930s. It was later found that DDT was toxic and it was killing local birds.

A Toxic Pesticide

DDT was sprayed onto crops and trees to kill insects. For many years DDT was thought to be harmless to larger animals, including humans. In the late 1950s, researchers began to realize that DDT was killing birds. They noticed that many robins were dying in areas where trees had been sprayed to kill elm bark beetles. Researchers discovered that earthworms were absorbing DDT and the robins that ate them were being poisoned.

Silent Spring

In 1962, scientist Rachel Carson alerted the public to the dangers of DDT in her book *Silent Spring*. She warned that if the "rain of chemicals" did not stop, Earth's ecosystems would be severely damaged. Pesticide manufacturers argued that the tiny amounts of DDT used to kill insects could not kill birds. However, further experiments showed that even small amounts of DDT could affect the reproduction patterns and the survival of many species, including birds. DDT was banned from general use in the United States in 1972.

Fast Fact
The United States Environmental Protection Agency estimates that more than 590,520 tons (600,000 t) of DDT was used in the United States.

Before DDT was known to be harmful, it was sprayed onto crops in the United States.

ISSUE 1

Toward a Sustainable Future: Identifying Toxic Organic Chemicals

If toxic organic chemicals are identified, people can replace them with safer alternatives. Scientists are researching safe levels of exposure to toxic organic chemicals and countries are banning chemicals that have proven to be toxic.

Safe Exposure to Toxic Organic Chemicals

Small amounts of some toxic organic chemicals may not be dangerous to humans. Researchers are testing many chemicals to see if they can be used at safe levels, or in amounts that will not be toxic to humans. Researchers measure the effects of these chemicals over time to make sure that there are no delayed effects, such as cancer.

The International POPs Elimination Network

The International POPs Elimination Network (IPEN) is a network of **non-governmental organizations** (NGOs) that formed in 1998. IPEN is working to remove **persistent organic pollutants** (POPs) across the globe. The network includes more than 400 public health, environmental, and consumer groups from 65 countries. They are implementing a treaty called the Stockholm Convention on POPs in countries around the world.

Fast Fact
Research shows that exposure to two or more toxic chemicals can increase the risk of disease. Smokers who have been exposed to **asbestos** are more likely to develop a type of cancer called mesothelioma.

Researchers test many different toxic chemicals to identify if they can be used at safe levels.

IPEN researchers found that free-range chicken eggs in many countries had high levels of toxic organic chemicals.

CASE STUDY
Toxic Chicken Eggs

In 2005, IPEN found that many free-range chicken eggs contained dangerous levels of toxic chemicals. This was the first time such extensive research about the presence of these chemicals had been carried out.

IPEN Research

IPEN researchers analyzed chicken eggs in 17 countries across the globe. The eggs were collected near waste incinerators, cement kilns, waste dumps, and chemical production plants. Researchers found that:

- every egg they analyzed contained **flame-retardants**
- seventy percent of the eggs also contained **dioxins**
- sixty percent of the eggs contained PCB levels above European Union limits

Importance of Measuring POPs

IPEN's research into free-range eggs shows that programs to measure the effects of POPs are vital. The director of the World Wildlife Fund's DetoX campaign, Karl Wagner, has said, "When even free-range eggs contain such high levels of chemicals, this shows that chemical contamination is totally pervasive. We must prevent the use of harmful chemicals. Without it we will not be able to trust the food we eat."

Fast Fact
When IPEN analyzed chicken eggs from a site in Egypt, they found the highest levels of dioxins ever measured in chicken eggs.

ISSUE 1

ISSUE 2

Lead Poisoning

Lead is a toxic metal that is used in many products all over the globe. People who have been exposed to even small amounts of lead over a long period of time can become sick or die as a result of the exposure.

Uses of Lead

Lead has been used in plumbing pipes for thousands of years because it does not **corrode**. Today, lead is also used in:

- gasoline
- batteries
- lead paint
- some cosmetics

Small amounts of lead are also used in the manufacture of many electronic devices, such as computers.

Fast Fact
The scientific symbol for lead is Pb. It comes from the Latin name for plumber, *plumberium*, which means "worker of lead."

Toxic Effects of Lead

Lead can build up in the body over time and cause many toxic effects. Lead is poisonous if it is eaten or breathed in as dust, fumes, or vapor. Exposure to even small amounts of lead can cause health problems as it builds up in the liver, kidneys, brain, bones, and teeth over many years. High levels of lead in the body may cause headaches, stomach aches, loss of energy, behavioral problems, and learning disabilities.

Lead pipes are still found in many parts of the world.

In China, many families collect electronic waste to recover toxic metals such as lead.

CASE STUDY
Recovering Lead in China

Lead from electronic waste, or e-waste, is creating a major health hazard in China. Today, many people who collect e-waste for recycling are at risk of developing lead poisoning.

Recovering Lead from E-waste

Many electronic devices, such as computers, contain lead, which can be recycled. When these devices no longer work, they become e-waste. E-waste is often shipped to China and other parts of Asia, where workers recover the lead for recycling. They often work under little supervision and without masks, gloves, or other safety equipment.

Families Recovering Lead

Members of many families in China and other **developing countries** also recover lead, mercury, and cadmium from scrap heaps of e-waste. Some families can earn enough money to pay for their needs by collecting these toxic metals. However, they can breathe in toxic dust, fumes, and vapor. Over time, these toxic chemicals can build up in their bodies and cause serious health problems.

Fast Fact
Plastics in e-waste are often burnt to recover lead and other metals. When these plastics are burnt, large amounts of dioxins are released into the atmosphere.

ISSUE 2

ISSUE 2

Toward a Sustainable Future: Preventing Lead Poisoning

Awareness programs and tight regulations governing the use of lead are now helping to reduce the impacts of lead poisoning.

Regulation of Lead

Many countries have regulations in place that govern the use of lead. Since the 1980s, several countries have banned or phased out products containing lead. According to the Center for Disease Control and Prevention, the number of children in the United States with high levels of lead in their blood dropped from 13.5 million in 1978 to 310,000 in 2002.

Limiting Lead Exposure

Programs such as the United States Environment Protection Agency's Lead Awareness Program are helping people to recognize the sources of lead in their environment and learn how to minimize their exposure to it. The main sources of lead exposure for children in **developed countries** are lead-based paint and lead-contaminated soil or dust.

Some ways to limit lead exposure are:
- using unleaded gasoline instead of leaded gasoline
- disposing of any lead-based paints safely
- washing hands after touching lead-contaminated soil
- cleaning dusty areas to prevent lead dust from building up
- avoiding toys, crayons, and cosmetics that may contain lead.

Unleaded gasoline has replaced leaded gasoline in most developed countries.

Fast Fact
It is illegal to use lead **solder** or lead pipes to transport drinking water in many countries.

In 2007, this toy was withdrawn from sale because it contained dangerously high levels of lead.

CASE STUDY
Removing Lead Toys in the United States

In the United States, many toys containing lead have been subject to **voluntary recalls**. Regulations are now in place to prevent the sale of products containing dangerously high levels of lead.

Voluntary Recalls

Many companies have launched voluntary recalls of products once the products have proved to be dangerous. In July 2004, some 150 million pieces of toy jewelry were recalled from vending machines when it was found that the jewelry contained dangerously high levels of lead. In 2006, some 300,000 charm bracelets containing lead were recalled after a child died from eating one. Other products that have been recalled include crayons, chalk, and clothing.

Banning Lead Toys

Today, many organizations are working to inform people about lead toys and ban them from sale. The Consumer Product Safety Commission publishes lists of products that are recalled because they contain dangerous levels of lead. In August 2008, the government introduced the *Children's Product Safety Act* to ban the sale of toys containing lead.

Fast Fact
Children can absorb lead more easily than adults, so they are at greater risk of lead poisoning.

Mercury Poisoning

ISSUE 3

Many people have been exposed to mercury in the environment and are suffering from the effects of mercury poisoning.

Uses of Mercury

Mercury is a valuable metal that is used in:

- paint
- fillings for teeth
- pipes and plumbing
- air conditioner filters
- cosmetics and jewelry
- some types of gasoline
- some processed foods
- some medicines and thermometers

Toxic Effects of Mercury

The toxic effects of mercury have been noted for thousands of years. Mercury poisoning is sometimes known as "mad-hatters disease." This is because a mercury-based solution was used to turn fur into felt for hats in the 1800s. People who made hats, called "hatters," breathed in mercury fumes and many developed mercury poisoning.

Mercury poisoning usually affects the nervous system and kidneys. Symptoms of mercury poisoning can also include loss of coordination, slurred speech, loosening teeth, memory loss, depression, anxiety, and trembling, or "hatter's shakes."

Fast Fact
The chemical symbol for mercury is Hg. Mercury is also the name of one of the planets in our solar system. It was named after the Roman god Mercury.

Many people get mercury poisoning from drinking water contaminated with mercury.

Fast Fact
The most common cause of mercury poisoning in people is eating contaminated seafood.

Today, Minamata Bay has safe levels of mercury and the fish have been declared safe to eat.

CASE STUDY
Minamata Disease

Between 1932 and 1968, Japan's Minamata Bay was polluted with mercury. Many people developed mercury poisoning, or Minamata disease, as a result of the pollution.

Mercury Pollution

Mercury was released into Minamata Bay from the Chisso Minamata Chemical Company's **acetaldehyde** chemical plant. It was eaten by tiny organisms that lived in the bay and their bodies converted it into methylmercury. Large fish and shellfish ate these organisms and methylmercury began to build up in their bodies. When humans ate these fish and shellfish, methylmercury was transferred into their bodies. Thousands of people who lived near Minamata Bay died of mercury poisoning and many babies were born with severe birth defects.

Timeline of the Development of Minamata Disease

1932	The Chisso Minamata Chemical Company begins producing acetaldehyde.
1953	First symptoms of mercury poisoning in local people.
1956	Minamata disease is discovered after patients die from unexplained symptoms at Minamata City Hospital.
1963	The Kumamoto University Study Group proves that Minamata disease is caused by methylmercury in Minamata Bay.
1968	The Chisso Minamata Chemical Company stops releasing mercury into Minamata Bay.
1972	The Minamata Disease Diagnosis Center and the Minamata Bay Pollution Protection Office are established.
1983–87	The Sludge Dredging Project removes toxic sludge from Minamata Bay.
1988	The Chisso Minamata Chemical Company is found guilty of releasing mercury into Minamata Bay.
1997	Fish in Minamata Bay are declared safe to eat.

ISSUE 3

ISSUE 3

Toward a Sustainable Future: Removing Mercury from the Environment

The number of people who die or are seriously affected by mercury poisoning each year can be reduced if mercury is removed from the environment.

Removing Mercury from Waterways

It is important to remove mercury from waterways so that it does not move up the food chain and poison humans. Small organisms often eat mercury that is released into waterways and their bodies transform it into methylmercury. Large fish, such as tuna, may eat these small organisms and become contaminated with methylmercury. Methylmercury can then move up the food chain and poison people who eat tuna and other large fish. It is difficult to remove mercury from waterways, but scientists are experimenting with using bacteria that can eat mercury.

The Zero Mercury Campaign

In 2005, the Ban Mercury Working Group launched the Zero Mercury Campaign. It aims to stop mercury from being released into the environment. The group was formed by 27 NGOs in 2002. Members of the group participate in the United Nations Environment Programme's Global Mercury Project. They are campaigning to close the world's largest mercury mine in Spain and they donate money to help those affected by mercury poisoning in developing countries.

Fast Fact
When mercury is released into waterways, it can become methylmercury, which is highly toxic.

These people are protesting against the use of mercury, which can pollute waterways.

Some fish that is sold in Australia has been tested to make sure that the levels of mercury are safe for people to consume.

CASE STUDY
Testing Fish for Mercury in Australia

In Australia, the number of people poisoned by mercury in the environment is getting smaller. This is due to regular testing of the mercury levels in fish and informing people about the dangers of eating contaminated fish.

Testing Fish for Mercury

Large fish are tested for mercury before they can be sold in Australia. The Australian Quarantine and Inspection Service (AQIS) tests the mercury levels in fish such as tuna, gemfish, and flake before they are imported into Australia. If the mercury levels are found to be too high, the fish is not allowed to be sold in Australia. Mercury testing is also carried out on fish caught in Australia.

Warning People about Mercury Poisoning

People have been warned to be careful of eating large amounts of fish such as tuna or flake, because they may contain high levels of methylmercury. In 2004, Food Standards Australia New Zealand (FSANZ) suggested there may be health risks for people who eat fish with high amounts of methylmercury. The warnings came after the Food and Agriculture Organization of the United Nations and the World Health Organization Expert Committee on Food Additives halved the tolerable weekly intake of methylmercury for pregnant women.

ISSUE 3

Fast Fact
The Stay Healthy, Stop Mercury campaign claims that a study showed that more than one in six women had mercury levels above safe standards.

Stockpiles of Toxic Chemicals

ISSUE 4

Toxic chemicals are often stored in stockpiles at factories or chemical plants. If toxic chemicals leak from these stockpiles, they can damage the environment and poison people who live nearby.

Storing Chemicals

Some industries use large amounts of dangerous chemicals to make products and they store these chemicals in stockpiles. When these chemicals are no longer needed, they are often transported by tanker to toxic waste disposal sites.

Chemical Spills

Stockpiles of toxic chemicals in factories or at chemical plants can leak and cause chemical spills or accidents. Leaks can contaminate nearby soil, air, or water supplies. They can also spread to rivers, lakes, and oceans. People, animals, and plants may then be poisoned by these toxic chemicals.

Toxic chemicals are often stored in stockpiles such as this.

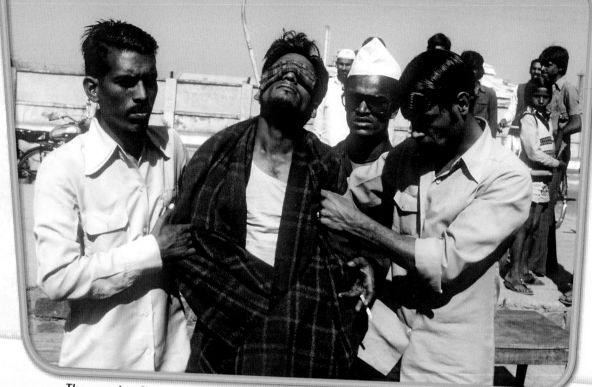

Thousands of people suffered from blindness following the Bhopal disaster.

CASE STUDY
The Bhopal Disaster in India

One of the worst toxic chemical leaks occurred on 3 December 1984, when a Union Carbide pesticide plant leaked clouds of deadly gas over the city of Bhopal, India.

A Deadly Accident

While most people in Bhopal slept, a lethal gas called methyl isocyanate leaked from the Union Carbide pesticide plant and spread over the city. The leak was caused by a series of mechanical faults and human mistakes at the plant. As a result of the leak, about 2,000 people died immediately and another 8,000 died later. Others suffered from blindness and breathing difficulties.

Fast Fact
Less than a year later, a Union Carbide plant in the United States leaked a toxic cloud of methyl isocyanate in the Kanawha Valley, West Virginia. Fortunately, no one was injured.

Lack of Preparation

Emergency services were unprepared for the disaster. Health officials were not told about chemicals at the Union Carbide factory, so they did not know about the potential for a chemical disaster. The people of Bhopal are still suffering from the effects of the disaster. Many babies have been born with birth defects and stunted growth, and this is believed to be as a result of the toxic chemical leak.

ISSUE 4

Toward a Sustainable Future: Safe Disposal of Toxic Chemicals

Toxic chemicals can be disposed of safely, so they do not need to be stockpiled. If the number of chemical stockpiles is reduced, there is less chance of chemical spills and contamination.

Safe Disposal Procedures

Many toxic chemicals can be disposed of safely if they are **incinerated** at high temperatures. Toxic chemicals need to be incinerated under supervision and with strict safety regulations in place.

Fewer Chemical Stockpiles

If more toxic chemicals are disposed of safely, fewer chemicals will need to be stockpiled in toxic waste disposal sites. Often, developed countries pay to send stockpiles of toxic waste to developing countries. This practice could be stopped to ensure that toxic chemicals are dealt with safely, not simply transported to create problems elsewhere on Earth.

Fast Fact
Pesticides need to be incinerated at temperatures of at least 1,652°F (900°C) to ensure that no dangerous gases are emitted.

Many toxic chemicals can be disposed of safely in a special waste incinerator.

Soil that has been contaminated by leaking stockpiles must be removed and disposed of safely.

CASE STUDY
The Africa Stockpiles Program

The Africa Stockpiles Program (ASP) aims to clean up and dispose of toxic pesticide stockpiles all over Africa.

> **Fast Fact**
> In the future, laws could be introduced to force companies to take responsibility for safely disposing of any toxic chemicals they use.

Stockpiles of Pesticides

An estimated 49,210 tons (50,000 t) of banned, contaminated, or expired pesticides are stored in stockpiles across Africa. Many stockpiled pesticides are in rusting drums inside rundown buildings, covered in plastic, in the outdoors, or buried under the ground. They often contaminate soils, food, water supplies, and air. These pesticides are also responsible for the deaths of about one million people every year.

Safe Disposal of Stockpiles

The ASP is locating pesticide stockpiles across Africa and sending them to Europe for safe disposal. The chemicals go to countries such as Wales and Finland, where there are special incinerators to dispose of toxic chemicals safely. The ASP is especially trying to locate ageing pesticide stockpiles. Pesticides are used to kill locusts and malaria-carrying mosquitoes, and the chemicals are stockpiled in preparation for future pest invasions. Many of these pesticides have been stockpiled for more than 40 years.

ISSUE 4

ISSUE 5

Toxic Chemicals at Home

Most people come into contact with toxic chemicals every day. Exposure to toxic chemicals can cause serious health problems.

Sources of Toxic Chemicals

Toxic chemicals are found in many products around the home, including:
- laundry products, such as laundry detergents, drain cleaners, and antibacterial cleaners
- bathroom products, such as air fresheners, toilet cleaners, and bleached toilet paper
- kitchen products, such as dishwashing liquids, oven cleaners, and some canned food products
- household cleaning products, such as furniture polishes, carpet shampoos, and mold and mildew cleaners
- cosmetics, such as hair dye and makeup

Fast Fact
Results from a survey across 12 European countries found a total of 73 **synthetic** toxic chemicals in the blood of 13 families.

Exposure to Toxic Chemicals

Exposure to toxic chemicals can cause:
- cancer
- damage to the nervous system
- birth defects in unborn babies

Some chemicals may not be classified as toxic, but they can still cause severe health problems if people are exposed to large amounts. The dangers of other chemicals may not yet be known.

Toxic chemicals are found in many common household cleaners.

24

Some common household paints and varnishes contain toxic chemicals.

CASE STUDY
Toxic Household Chemicals in Australia

The Child Accident Prevention Foundation of Australia reports that around 50 children per week are admitted to hospital with poisoning caused by exposure to toxic chemicals at home.

Poisoning Accidents and Deaths

Each year, 140,000 calls are made to Poisons Information Centers by concerned parents or guardians across Australia. About 3,500 children under five years of age are admitted to hospitals each year due to poisoning. Five to 10 of these children die each year.

Sources of Poisoning

Exposure to toxic chemicals is responsible for 28 percent of poisonings among children. Household products that are commonly involved in these occurrences include cleaning products, pesticides, and fertilizers. Soaps, shampoos, deodorants, perfumes, medicines, cough syrups, prescription drugs, and vitamins have also caused poisonings among children.

Fast Fact
Many household products that are not listed as toxic could still poison children, especially if children are exposed to large amounts of these products.

ISSUE 5

Toward a Sustainable Future: Reducing Exposure to Toxic Chemicals

People can reduce their exposure to toxic chemicals by disposing of dangerous products safely. They can also replace dangerous products with non-toxic alternatives.

Disposing of Dangerous Products

Following the directions on dangerous products to dispose of them safely will help to prevent toxic chemicals from polluting the environment. Some products, such as paint and oil, should not be poured down household sinks. Other products, such as electronic devices and batteries, should not be thrown in garbage cans.

Finding Non-toxic Alternatives

Dangerous products used around the home can often be replaced with non-toxic alternatives. Many household cleaners and fertilizers are made using chemicals that are not toxic to humans. These alternatives are safer and, often, they are also better for the environment. Using these products means that fewer toxic chemicals are washed down household sinks and drains.

Fast Fact
In the United States, most people are poisoned at home when they take the wrong amount of medicine, or take the wrong medicine.

Some bathroom products contain ingredients that are not toxic to humans.

Fast Fact
Mixing some common household cleaners can cause a chemical reaction that releases toxic gases.

Safety campaigns advise people to wear protective goggles and gloves when handling dangerous chemicals.

CASE STUDY
Fire Brigade Safety Campaigns

Fire brigades learn how to handle toxic chemicals safely. They also give advice to people on how to handle these chemicals safely.

Safe Handling of Toxic Chemicals

Fire brigades are often called when people are poisoned by toxic chemicals. Members of these fire brigades must know how to handle toxic chemicals safely to prevent chemical spills. Many fire brigades also run safety campaigns about toxic chemicals and provide people with information about how to handle chemicals safely. This helps people to understand the risks in handling toxic chemicals and how to avoid them.

Tips for Handling Toxic Chemicals Safely

Many fire brigades offer tips for handling toxic chemicals safely. These include:
- reading labels to make sure that you understand how to use the product safely
- following all the safety instructions listed on the product
- wearing protective clothing, such as gloves, hats, and masks, if necessary
- cleaning equipment after use, even if it will be thrown away
- storing chemicals in their original containers
- locking chemicals away in a safe place when they are not being used
- disposing of chemicals safely

ISSUE 5

What Can You Do? Choose Non-toxic Products

Many non-toxic products can be used in place of products that contain toxic chemicals.

Look for Non-toxic Products

There are many non-toxic products that can be used in place of dangerous products. You can look for non-toxic products in supermarkets and retail stores. You can tell your family about these non-toxic products and encourage them to buy or use the products.

Some non-toxic alternatives to dangerous household products are listed in the table below.

Fast Fact
If dangerous chemicals are poured down the drain or toilet, they will likely end up rivers, streams or oceans.

Area of Use	Dangerous Products	Non-toxic Products
Bathroom	• cosmetics, soaps, shampoos, and hair dyes with synthetic scents	• cosmetics, soaps, shampoos and hair dyes with no synthetic scents
Kitchen	• oven cleaners and floor cleaners with synthetic scents • cling wrap	• oven cleaners and floor cleaners with no synthetic scents • glass storage containers
Laundry	• laundry detergents with synthetic scents • chemical-based stain removers	• laundry detergents with no synthetic scents • natural stain removers that can be made at home
Living room	• chemical-based air fresheners • chemical-based carpet cleaners	• fresh flowers • natural oils • beeswax candles

Make Green Cleaners

You and your family can make environmentally friendly products, or green cleaners, at home. These products are better for you and better for the environment. Below are some recipes to make green cleaners at home.

Clothing Stain Removers

Mix equal amounts of soda water, cold water, and lemon juice. Apply this mixture to stained clothing before it is washed.

Carpet Stain Removers

Sprinkle salt and a small amount of lemon juice onto stained carpets. Leave the mixture for a few hours, then wipe clean.

Window Cleaners

Add one cup of vinegar to a bucket full of water. Use the mixture to clean the windows, using old newspaper as a cleaning cloth.

Air Fresheners

Mix a few drops of your favorite natural oil with warm water. Put this mixture into a bowl.

Clean your windows with old newspapers and a green cleaner made from vinegar and water.

Toward a Sustainable Future

Well, I hope you now see that if you take up my challenge your world will be a better place. There are many ways to work toward a sustainable future. Imagine it ... a world with:
- decreasing rates of global warming
- protected ecosystems for all living things
- renewable fuel for most forms of transportation
- sustainable city development
- low risks of exposure to toxic substances
- a safe and reliable water supply for all

This is what you can achieve if you work together with my natural systems.

We must work together to live sustainably. That will mean a better environment and a better life for all living things on Earth, now and in the future.

Web Sites

For further information on toxic waste, visit these websites:
- World Wildlife Fund toxic blaster game
  www.panda.org/games/toxicblaster/game.html
- United States Environmental Protection Agency
  http://www.epa.gov/kidshometour/
- National Library of Medicine Tox Town
  http://www.nlm.nih.gov/pubs/factsheets/ToxTown.html

Glossary

acetaldehyde
an organic chemical used in a range of products, including perfumes, dyes, plastics, and synthetic rubber

asbestos
a strong, fire-resistant material that can cause cancer

chemical residues
small amounts of chemicals that are left in the environment after toxic waste breaks down

contaminate
add dangerous or unwanted substances to the environment

corrode
gradually rust or waste away if exposed to oxygen

DDT
dichlorodiphenyltrichloroethane; a toxic chemical used on plants in the United States during the 1930s

developed countries
countries with industrial development, a strong economy, and a high standard of living

developing countries
countries with less developed industry and a lower standard of living

dioxins
a group of chemical compounds

exposed
coming into contact with a toxic substance without protection

flame-retardants
chemicals that are sprayed on materials to reduce burning

global warming
a rise in average temperatures on Earth

incinerated
burned at very high temperatures until only ashes are left

inorganic chemicals
chemicals that come from minerals, including metals

non-governmental organizations (NGOs)
non-profit groups not run by governments

nonrenewable
a resource that is limited in supply and cannot be replaced once it runs out

organic chemicals
complex chemicals formed by living things and containing carbon

persistent organic pollutants
dangerous chemicals that remain in the environment for a long time

pesticides
substances that are sprayed on plants to kill insects

poisons
substances that cause illness or death to living things

radioactive materials
materials that give off radiation, which can cause sickness in living things

solder
a mixture of metals, usually lead and tin, which is melted and used to join metal pipes together

synthetic
made by people

voluntary recalls
asking people to return dangerous products to the companies that made them

Index

A
acetaldehyde 17
Africa Stockpiles Program 23
air pollution 5, 13
asbestos 10

B
Ban Mercury Working Group 18
Bhopal, India 6, 21
birds 6, 9, 11
brain damage 8

C
cancer 8, 10, 24
Carson, Rachel 9
chemical spills 6, 20, 21, 22, 27
chemical stockpiles 6, 20, 23
China 7, 13
Chisso Minamata Chemical Company 17
cleaning products 6, 7, 24, 25, 25, 26, 27

D
DDT 6, 8, 9
DetoX campaign 11
dioxins 11, 13
dishwashing liquids 24, 28

E
electronic waste 7, 12, 13, 26
environmental disease 8

F
fish 17, 18, 19
flame retardants 11
Food Standards Australia New Zealand (FSANZ) 19

G
green cleaners 29

H
health problems 5, 13, 24
household toxic chemicals 6, 7, 24–28

I
inorganic chemicals 5
International POPs Elimination Network (IPEN) 10, 11

L
lead pipes 12, 14
lead poisoning 6, 7, 12–15
lead toys 14, 15

M
mad-hatters disease 16
medicines 5, 26
mercury poisoning 6, 7, 13, 16, 17, 18, 19
methyl isocyanate 21
methylmercury 17, 18, 19
Minamata Bay, Japan 6, 17
Minamata disease 17

N
non-toxic alternatives 26, 28

O
organic chemicals 5, 6, 8–11

P
PCBs 8, 11
persistent organic pollutants (POPs) 10, 11
pesticides 5, 6, 8 9, 21, 22, 23, 25

R
radioactive waste 5
regulations 9, 14, 15, 19

S
safe disposal of chemicals 22, 23, 27
safe handling of chemicals 27
Stay Healthy, Stop Mercury campaign 19
Stockholm Convention on POPs 10

T
toxic chicken eggs 11
toxic waste disposal sites 20, 22

U
Union Carbide chemical plant 21
unleaded gasoline 14

V
vinyl chloride 8
voluntary recalls 15

W
waste incinerators 22, 23

Z
Zero Mercury Campaign 18